On the Energy Structure of Natural Vegetation

In search for community governance rules

Briefly about the book …

Vegetation Science meets quantum theory in the energy-based entropy model of this book. The model is based on Max Planck's postulate that potential energy and entropy are interchangeable parameters in resonator complexes. What is a typical outcome of the model in vegetation studies? The model hands users a set of entropy estimates and probabilities based on which the strength and uniqueness of a metacommunity's energy structure can be characterised in comparative terms.

To Márta

ON THE ENERGY STRUCTURE OF NATURAL VEGETATION

In search for community governance rules

László Orlóci, FRSC

Visiting Professor, Laboratory of Plant Quantitative Ecology,
Universidade Federal do Rio Grande do Sul, Porto Alegre, Brazil

SCADA – London, Canada

Scada
Publishing

Refer to this book as

Orlóci, L. 2013. On the energy structure of natural vegetation. In search for community governance rules. SCADA Publishing, London, Canada. Enlarged Online Edition: https://createspace.com/4153484

Look for

Orlóci, L. 2012. Self-organisation and Mediated Transience in Plant Communities. SCADA Publishing, London, Canada. Enlarged Online Edition: https://createspace.com/3585127

Orlóci, L. 2011. Statistical Ecology. The quantitative exploration of nature to reveal the unexpected. SCADA Publishing, London, Canada. Online Edition: https://createspace.com/3476529

Orlóci. L. 2011. Problem flexible computing in statistical ecology. SCADA Publishing, London, Canada. Online Edition: https://www.createspace.com/3574792

Orlóci, L. 2012. Statistical multiscaling in dynamic ecology. Probing the long-term vegetation process for patterns of parameter oscillations. SCADA Publishing, London, Canada. Online Edition: https://createspace.com/3830594

ISBN-13: **978-1482319378**
ISBN-10: **1482319373**

2013 enlarged edition

*"Alle Gestalten sind ähnlich, und keine gleichetder andern;
und so deutet das Chor auf ein geheimes Gesetz, ..."*

Johann Wolfgang von Goethe (1798): "Die Metamorphose der Pflanzen "

www.udoklinger.de/Deutsch/Goethe/quelle.htm

In free translation:

"All [plant] forms are similar, but none like any other;
and so the ensemble interprets a hidden law ..."

Credit for the Coquihalla data set goes to Márta Mihály BSF, DFE.

Abstract ...
The search for governance rules, which are assembly/disassembly rules in natural plant communities, has been responsible for the ascent of vegetation studies into the realm of modern science. The conceptual underpinnings evolved over time into their present complex state by necessity of the realisation that the vegetation process is energized by convoluted effects issuing from the processes of phylogeny, environmental mediation, and unavoidable chance. The isolation of the individual effects became a central research problem. Current models parameterised by the specific variance, covariance and correlation provide a practical solution. Further clarification of the concepts, and development of statistical tools to open the way to parameterisation of an energy-based entropy model, and illustration how model works in practice, are the book's main objectives.

Keywords: Assembly; Disassembly; Coquihalla floodplain; Diversity; Governance rules; Entropy; Environment; Evolution; Hierarchical relevé; Plant systematics; Phylogenetic tree; Potential energy; Resonator complex; Succession; Vegetation.

TABLE OF CONTENTS

SYNOPSIS

The Science of Community Ecology is heavy with assumptions concerning governance in the natural community assembly (disassembly) process. When interconnected within a consistent analytical model, the assumptions become the basis of a general scientific theory.[1]

There are three assumptions on which I focus attention in this Essay:

1. The long-term evolutionary process. Evolution endows the com-

---

[1] I refer to two recent publications for examples and further references. Let these speak for themselves:

Pillar, V.D. and L.S. Duarte. 2010. A framework for metacommunity analysis of phylogenetic structure. Ecology Letters 13: 587–596.

Orlóci, L. 2012. Self-organisation and Mediated Transience in Plant Communities. SCADA, London, Canada. Enlarged Online Edition: https://createspace.com/3585127.

munity elements, the plant taxa, with the functionality of self-organisation.

2. Current environmental mediation. This sets limits on self-organization by triggering compositional changes. We refer to this process as community transience.

3. Chance. The determinism of evolution and environmental mediation are braided with chance effects. The process in such a convolution is never completely deterministic or completely ruled by chance.

I mention three models:

1. The model of Pillar and Durante (2010). The model parameter is a partial correlation without an explicit hierarchical structure.

2. My 2012 model. This works with an explicit hierarchical parameter structure (specific variance, covariance) which allows multiscaling to pinpoint the level in the evolutionary tree (evolutionary plant systematics) on which the parameter structure is sharpest and environmental mediation is maximal.

3. The energy-based entropy model. This Essay rewrites the model in the terms of Max Planck's[2] quantum theoretical energy-based entropy. It takes advantage of the actual fact that energy and entropy are interchangeable parameters. The model itself is hierarchical and probabilistic. The hierarchical structure allows incorporation of components for the phylogenetic history of the taxa in tandem with current environmental mediation. Because the model hands us probabilities, the apportionment of the energy-based entropy between the model components is readily seen in comparable terms.

New relationships revealed by model illuminate the intrinsic (hidden) rules. Plank's model is unique in that respect that we are required to think of energy not as a continuous wave, but as a stream of discrete, countable energy packets, the quanta. This translates

---

[2] http://en.wikipedia.org/wiki/Max_Planck

into operational Vegetation Science[3] in a simple way: when we measure a vegetation state we are in fact counting energy units that have been put to work to attain the state.

Naturally, we have to be aware that energy is completely object (taxon in this case) neutral. Therefore the model is entirely in a probabilistic diversity space steps removed from the taxonomic identities of the plants. The analysis is comparative, but the comparisons are now using the energy-based entropy level of the plant community. This implies that for the ecologist interested in taxon-to-taxon comparisons the model falls one step beyond the rules he or she may be attempting to discover.

---

[3] Alternative designations for Vegetation Science: Phytosociology, Plant Ecology, Plant Community Ecology, Vegetation Ecology. The same in Vegetation Science as with anything else in ecological concerning the vegetation process, the approach is Poorean successive approximation. Therefore, the limit of effort is set by the researchers thrive for precision in logic and physical recording under strict time and monetary restrictions.

RETROSPECT

I am adapting this section from my companion book of 2012, which presented the specific variance model. As I have said earlier: a re-ignited interest in finding the rules of governance in plant community assembly/disassembly heralds no novel undertaking by me or by others of my contemporaries. The search for the rules has been in fact the principle incentive in the development of plant ecology as a science.

The early period in this development is noted for the publication of the elegiac poem "Die Metamorphose der Pflanzen" in 1798 by the poet naturalist Johann Wolfgang von Goethe. The poem itself speaks of an ensemble of plant forms which interpret a hidden law by the diversity of its membership. Speaking of a hidden law in connection with an assemblage, such as a diverse well-delineated natural plant community, and linking the behaviour of this assemblage to an intrinsic law, must have appeared to naturalists of that era, tormented by great revolutions, to be by itself quite revolutionary. The more so since science's plant community concept at that time had yet to progress beyond the aesthetic plant geographic view which Sukopp (1987) finds dominant in Humboldt and Bonplan's "Essaie sur la Géographie des Plantes" of 1805.

Goethe's ideas regarding the morphology and functioning of plants, and his clearly expressed notion of plant assemblages discharging functions like an orchestra in the presence of a hidden law could not possibly go unnoticed among 19th Century naturalists. It may not be a far fledged proposition therefore that Goethe's ideas are precursor to Kerner von Marilaun's doctrine of plant community development. Kerner does quote Goethe in his seminal 1863 book "Das Pflanzenleben der Donauländer".

Kernerian plant community development is a site related (point) assembly/disassembly process. It involves compositional transitions in time mediated by action-reaction feedback. When ecologists speak of facilitation, the motor of community transience, they identify the Kernerian mechanism -- with or without any awareness of Kerner's work.

The central core of Kerner's doctrine is the proposition of a mediated tendency in the plant community's functioning to change the environment at the potential cost of its own transience. How did Kerner come to the idea of temporal compositional transitions undertaken by the plant community in situ? His approach is a case of space-for-time substitution[4]. It could have happened like this:

1. General reconnaissance of highly dynamic alluvial sites[5] reveals coincidental patterns of plant community types and nominal substrate age.

2. Space (the land pattern of community types) linked to time (sediment age), the conclusion of plant community development by way of temporal compositional transitions in situ is in hands.

I should mention as corollary to the space-for-time substitution

---

[4] Wildi and Schütz (2000) describe a complex technique. Fractional time series are taken from permanent plots and spliced into a single chain. This chain is used as surrogate in the prediction of succession over the time span. The permanence of the same kind of climate is assumed.

[5] I refer here to Kerner's Achen Lake example from Tirol.

approach that the developmental series is defined for community types which happened to be observed in the studied site. Community types potentially present but not observed in the sites are left out.

I leave it to the ambitious student reader to fill the gap left by what has been said in retrospect and what is going to be discussed in detail in the sequel.

PRELIMINARIES

Fundamental truths

We know that nothing can be done without energy. We know from Max Planck (1901) that energy comes in countable units. In fact Planck's numerical example uses the quantum idea in a context which suggests direct applicability in Vegetation Science. This requires the interpretation of physical measurements as counts of energy units, the quanta. I pursue this property and construct a methodology around it. I argue that because the conditions which Max Planck specifies are universal, the energy-based entropy equation of Max Planck is suitable to re-parameterise my 2012 model which used the specific variance.

The context of my original model is changed by being embedded within an entropy based reference system. The methodology had to be focussed on data total and the numbers of data elements contributing to the total, not on the data elements *per se*. This may alarm the ecologist believing that by so doing information is lost.

This is true, but an ingenious solution is gained for making inferences from entropy about its parameter dual, potential energy.

By its very nature the energy-based entropy model is stochastic. It serves the objective of tracing in quantum theoretical terms the historic energy path though the levels of the phylogenetic system, or equivalently through the nodes of the evolutionary tree, and the loci on a gradient of environmental mediation. In this way I can interconnect two major mechanisms responsible for governance in the plant community assembly/disassembly process in energy terms in a probabilistic context.

The hierarchical relevé

Figure 1 portrays three dendrograms R1, R2 and R3. These are proxy mappings of the phylogenetic tree reconstructed by evolutionary plant systematists. To an ecologist my dendrograms represent *hierarchical relevés* which describe metacommunities in nature[6].

Any node of the phylogenetic tree marks the position where a higher level taxa splits into lower level taxa, say Order into Families, Families into Genera and so on. Figure 1 presents a fictitious case. The basic scheme used in real cases comes directly from the manual which experts use for plant identification. These are assumed to be consistent with the principles of modern phylogenetics.[7]

---

[6] To the data analyst R1, R2 and R3 are hierarchical record sets of a special kind. The top level incorporates 2 nodes which branch into 4 nodes which in turn branch into 6 nodes. In the examples to be presented the dendrogram levels correspond to taxa of increasingly higher systematic status (species, genus, family, order, class). The branching pattern is the same in R1, R2 and R3. The baseline data sets [5 7 3 2 2 4], [7 9 9 1 8 8], [4 9 3 9 7 7] and upper level cumulants are different. The identical brunching pattern indicates identical sets of baseline taxa (a to f). The baseline data elements are indivisible. They can be pooled to form cumulants on the upper hierarchical levels.

[7] To answer arguments on the contrary in specific cases is beyond my intention in this essay. I leave that discussion to the experts of evolutionary plant systematics.

On the energy-based entropy structure of natural vegetation

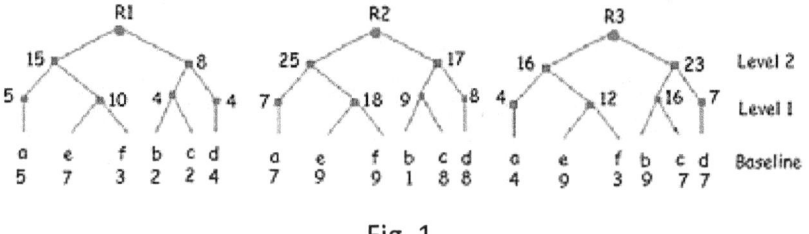

Fig. 1

ENERGY-BASED ENTROPY

In general

As we enter the reference space of *energy-based entropy* -- which most of us knew in its roughest outlines from high school physics, but forgot -- our first act is to strip the hierarchical relevé of its ecological label and refer to it as a *complex*. We rename the taxa as *resonators* and replace the base line code (a, b, c, d, e, f) by integer numbers (1 to 6 in figure 1). On the 0 or baseline level we have for the hierarchical relevés in Figure 1:

	Resonator						
	1	2	3	4	5	6	Total
Complex							
R1	5	7	3	2	2	4	23
R2	7	9	9	1	8	8	42
R3	4	9	3	9	7	7	39
Total	16	25	15	12	17	19	104

On the energy-based entropy structure of natural vegetation

We take each number in the body of the table as a count of energy units. For example the number 9 reads "nine energy units", 23 as "twenty three energy units", and so forth. If we denote by ε the energy of one unit, the total energy is 104ε in the 6 x 3 complex.

Theory

The details follow closely Max Planck's 1901 paper. My symbols are not always the same as the symbols in the original paper.

Now the basic definitions:

$U_n = nU$ total energy in a complex of n resonators

$U = \dfrac{U_n}{n}$ energy of one resonator

T sum of the number of energy units

ε unit energy (one quantum)

$U_n = T\varepsilon$ energy of the n-resonator complex

$H_n = nH$ entropy of the complex

$H_n = k \ln W + constant$ energy-based entropy equation

H entropy of a resonator

The 1901 paper provides the proof which shows that H is in fact proportional to U by way of probability $1/W$. Symbol W denotes the number of all possible equally probable complexes with the same T, n and U_n. W is estimated by the combinatorial equation

$$C = \frac{(n+T-1)!}{(n-1)!T!}$$

Max Planck used Stirling's approximation replacing $n!$ by n^n to obtain

$$C = \frac{(n+T)^{n+T}}{n^n T^T}$$

The combinatorial given in these particular terms facilitated the algebra which connects H to U:

$$H_n = k \ \ln \ C = k[(n+T) \ \ln \ (n+K) - n \ \ln \ n - T \ln T]$$

$$= kn\left[\left(1+\frac{U}{\varepsilon}\right)\ln \ \left(1+\frac{U}{\varepsilon}\right)-\frac{U}{\varepsilon} \ \ln \ \frac{U}{\varepsilon}\right]$$

and further $H = k\left[\left(1+\dfrac{U}{\varepsilon}\right)\ln \ \left(1+\dfrac{U}{\varepsilon}\right)-\dfrac{U}{\varepsilon} \ \ln \ \dfrac{U}{\varepsilon}\right]$. Since H is the en-

tropy of one resonator, H_n in any complex R1, R2 or R3 is proportional to the energy U_n. Further, since R1, R2 and R3 are natural complexes, it is not far fledged to assume to satisfy Planck's conditions that these complexes exists under the *rule of chance* and all complexes are *equiprobable*.

What is energy?

The attentive reader may have already wanted me to say "what do I mean by energy"? I refer to the answer which the Nobel Laureate physicist Richard Feynman gives in his famous lectures of 1964 to undergraduates[8]: "... we have no knowledge what energy is"[9,10]. This is not to say that there is such a thing as energy which we all know first-hand from experience with the manifestations when it is put to work.

In our space/time frame of reference we speak of H_n as the total entropy of the relevé (the complex), H as the entropy of a taxon (unidentified), and so forth. By the same token we may speak of the total energy of a relevé and the energy of a taxon.

I suggest that the entropy we are considering issues from three sources. One is directly a consequence of the evolutionary proliferation of taxa in the manner of the phylogenetic tree, the other is

---

[8] Most students in class, according to hearsay, did not really benefit from Feyman's lectures. The experiment of an expert from the highest level teaching an introductory course had to be reconsidered.

[9] http://en.wikipedia.org/wiki/Energy -- Richard Feynman, in The Feynman Lectures on Physics (1964) Volume I, 4-1

[10] http://en.wikipedia.org/wiki/The_Feynman_Lectures_on_Physics

owing to environmental mediation, and the third to chance. It follows that the systematic position on which the taxa are identified (class, order, etc.), the loci of the relevé on the environmental gradient, and random effects do matter in our search for the principles of governance.

The process of evolution cannot be reversed in Nature. Yet we have to look back to an earlier node on the phylogenetic tree when we characterise the source of the entropy we are dealing with. Note the long time-scale involved in this. When we look back we see a rise in the value of H and a decline in the number of taxa. Physics' technical term for the associated U is *potential energy*. It is in fact the energy spent to increase the richness of taxa. This has its analogue in physics when an object increases its potential energy in proportion of the work needed to lift it to a higher point in a gravity field and spends it when allowed to drop to a lower point. The same considerations apply to the potential energy that is put to work in the process of Kernerian facilitation which in part is powering environmental mediation.

These bring us to the next question.

Can we measure U?

The answer is yes, but we have to put it to work and measure what is being produced by experiment. How does this connect to cover/abundance estimates? - or as a matter of fact, how does it connect to any other quantitative measure of a taxon's mass? Perhaps, the measurement of heat energy released when samples of harvested biomass are ignited would hand us one empirical constant.

Why do I want to determine a constant locally when there is Max Planck's universal constant? Intuition is telling me that when we are transplanting ideas about energy-based entropy into a field in which as far as I know they were not tested, it would be interesting to verify the energy-based entropy's energy predictions.

ENTROPY NUMERICS

We now turn to the R1, R2 and R3 complexes and develop the analytical technique which we need in the interrogation of their energy-based entropy states. We compute C and H_n based on T and n. Other entropy quantities are derived from H_n and n according to the definitions already given and as will be defined in the sequel.

The basic data set comes from Fig. 1:

R1 complex		R2 complex		R3 complex	
T	n	T	n	T	N
23	2	42	2	39	2
23	4	42	4	39	4
23	6	42	6	39	6

The first set of results:

Complex R1	C	H_n	C_{spec}	$H_{n\,spec}$	n	H_{spec}	P
Level 2	1063	6.97	1063	6.97	2	3.39	0.03
Level 1	82955	11.33	78	4.36	2	2.18	0.11
Baseline	2635689	14.78	32	3.46	2	1.73	0.18
			H_n	14.78		2.46	0.09

On the energy-based entropy structure of natural vegetation

Since our complexes are hierarchical relevés we have made provision for the dependence of branching at a lower level on branching of the next higher level. We pass from the C and H quantities to their specific forms C_{spec} and H_{spec}. Regarding these we note:

Level 2

$C_{spec} = C = 1063.41$

$H_{n\ spec} = H_n = 6.97$

$H_{spec} = 6.97/2 = 3.48$

Level 1

$C_{spec} = 82954.78/1063.41$

$H_n = ln(78.01)$

$H_{spec} = 4.35/(4-2) = 2.18$

We note further that C can be excessively large. We bypass the computational problem by going directly to

$$H_n = (T+n)\ \ln\ (T+n) - T\ \ln\ T - n\ \ln\ n$$

Further, for any type of H a probability is defined by

$$P = e^{-H}$$

under the rule of chance.

How do we interpret the results? Some pointers:

1. The inverse of C_{spec} is a conditional probability. $H_{n\ spec}$ is the conditional entropy of the complex on a given dendrogram level's number of nodes n. The specific entropy of a taxon on that level is H_{spec}.

2. The total of the $H_{n\ spec}$ quantities is H_n for the hierarchical complex.

3. The potential energy U is proportional to the H values, total or partial.

4. The probability P is our measure a complex's or its partition's uniqueness in an inverse manner. The smaller is this probability the more likely that H is not a consequence of a fluke chance event in sampling.

Similar calculations such as the above are performed for R2 and R3 and summarised in the similarly arranged tables:

Complex R2	C	H_n	C_{spec}	H_{nspec}	n	H_{spec}	P
Level 2	3415	8.14	3415	8.14	2	4.07	0.02
Level 1	798271	13.59	234	5.45	2	2.73	0.07
Baseline	71482823	18.08	90	4.49	2	2.25	0.11
			H_n	18.08		3.01	0.05

Complex R3	C	Hn	C_{spec}	H_{nspec}	n	H_{spec}	P
Level 2	2955	7.99	2955	7.99	2	4.00	0.02
Level 1	601705	13.31	294	5.32	2	2.66	0.07
Baseline	47221083	17.67	78	4.36	2	2.18	0.11
			H_n	17.67		2.95	0.05

Guided by the example already explained, we can say with certainty that in absolute terms, and only in those terms, the total energy in R1 is lower than in R2. But saying this may not be sufficient. We have to consider the statistical sampling error and based on that draw into the interpretation of any H the idea of statistical uniqueness. P comes handy for the task.

We should consider the hierarchical relevés on hand as a random sample from a universe of equally probable hierarchical relevés we could take in a given site, and for that exact reason *a priori* not unique. Whether we continue holding this view will depend on our interpretation of P.

Note that the H_{spec} quantities 2.46 and 3.01 are proportional to effects which triggered the difference 3.01 - 2.46 = 0.55. The probability of an equally larger difference occurring by chance alone at the given n and T is equal to $e^{-0.55} = 0.05/0.09 = 0.58$. This is huge. We need no further comparisons to R3 to make our overall decision considering that the R1 and R2 entropy quantities are the most different in the sample. Taking all the noted properties into account, we conclude that the examined differences in energy-based entropy have no statistical uniqueness, *i.e.* they are well within any reasonable limits of random expectation.

This conclusion has serious implications when [R1 R2 R3] mark successive loci on a well-defined environmental gradient. In that case we would have to conclude that the environmental effect was too weak to bear on unique differences between the hierarchical relevé

On the energy-based entropy structure of natural vegetation complexes in energy-based entropy terms.

There are different directions to take at this junction in the analyses. We explore the following two:

The concatenated hierarchical relevé complex

We assume a well-defined elevation gradient traced by the R1, R2 and R3 loci in that sequence. What is the effect on energy-based entropy? Analysis of the concatenated baseline data and the cumulants hold the answer to this:

```
Cumulants by level for concatenated relevés
level 2: 15 8 25 17 16 23
level 1: 5 10 4 4 7 18 9 8 4 12 16 7
baseline: 5 7 3 2 2 4 7 9 9 1 8 8 4 9 3 9 7 7

level totals and record string lengths
 104   6
 104  12
 104  18

C* Hn H P by level across relevés
level 2 :-- 23.28563 3.8809383 2.0631457e-2
level 1 :-- 38.580929 3.2150774 4.0152226e-2
Baseline:-- 51.047222 2.8359568 5.8662371e-2
 *direct computation of C bypassed
```

What do these numbers tell about the elevation catena? We can draw up a table and examine the numbers:

		H_n	$H_{n\,spec}$	n	H_{spec}	P
Level 2	:--	23.29	23.29	6	3.88	0.02
Level 1	:--	38.58	15.30	6	2.55	0.08
Baseline	:--	51.05	12.47	6	2.08	0.13
		H_n	51.05	18	2.84	0.06

The H_{spec} values tell us that the concatenated hierarchical complexes make for a rather unique chain in environmental progression. Namely, 3.88, 2.55, 2.08 and 2.84 tell us that the catena on all hierarchal levels is unique.

Hierarchical relevé complex as a rectangular array

Having three baseline vectors, we have a rectangular array:

```
5 7 3 2 2 4
7 9 9 1 8 8
4 9 3 9 7 7
Relevé (row) totals
23 42 39
Taxon (column) totals
16 25 15 12 17 19
```

What could be a reason calling for analysis of such an arrangement of the relevé complexes? – typically, when environmental mediation is not on a catena but distributed over the landscape in some pattern. The rectangular array is the mapping of the response levels. The analysis follows what we usually do in canonical contingency table analysis, but at this time we use totals, and leave the individual cells untouched:

```
(A) C* Hn H P within columns
   col 1 -* 8.2870842 2.7623614 6.3142488e-2
   col 2 -- 9.5339938 3.1779979 4.1668996e-2
   col 3 -- 8.1101018 2.7033673 6.6979595e-2
   col 4 -- 7.5060364 2.5020121 .08192
   col 5 -- 8.4541818 2.8180606 5.9721656e-2
   col 6 -- 8.7627565 2.9209188 5.3884154e-2

(B) C* Hn H P within rows
   row 1: -- 14.784655 2.4641092 8.5084601e-2
   row 2: -- 18.084968 3.0141613 4.9086988e-2
   row 3: -- 17.670351 2.9450585 5.2598982e-2

(C) Main effect columns
   C* Hn H P
   -- 23.28563 3.8809383 2.0631457e-2
   Main effect row
   C* Hn H P
   -- 13.680195 4.560065 1.0461379e-2
   Joint effect:
   C* Hn H P
   -- 51.047222 2.8359568 5.8662371e-2
   Interaction
   -- 14.0814  1.5646   0.2092
   *Computation of C is bypassed in all cases.
```

Here we have some complexities which require explanations:

1. Consider the results in group (A). A typical subset comes about such as these:

On the energy-based entropy structure of natural vegetation

$H_n = (16+3) \ln (16+3) - 16 \ln 16 - 3 \ln 3 = 8.287$
$H = 8.287/3 = 2.762$ and
$P = 1/\exp(2.762 = 0.063$

These help us to evaluate uniqueness within columns or between columns. For example $P = 0.063$ indicates a rather unique case and $|2.762 - 3.178| = 0.416$ and $P = 1/\exp(0.416) = 0.660$ indicates a rather <u>not</u> unique case. This should set a pattern for the interpretation of the other values by readers.

2. In group (B) we see the results reappear from the analysis of the individual complexes (R1, R2, R3).

3. The group (C) holds the energy-based entropy of the rectangular array partitioned by main effects and joint effect. The difference 51.047 -23.286 -13.68 = 14.081 is the emergent interaction term. The associated H value is 14.081/9 =1.565. The probability that a value as large as 1.565 could arise by chance is $P = 1/\exp(1.565) = 0.209$, far too large to warrant declaration of uniqueness.

COQUIHALLA CASE STUDY

Survey site

This example uses the M. Mihály data set, a portion of which is presented in Table 1. The data set comes from a transect survey on the inner side of a major meander on the Coquihalla River floodplain in Hope, British Columbia (Fig. 2).[11] The end points of the transect line are marked on the map in Fig. 2. The site itself is no longer the same.

At the time of the survey (July 7-8, 1976) a natural high forest covered the site and the flow in the river was uncontrolled. Now the site is occupied in good part by a subdivision of Hope and the flow in the river is manipulated upstream.

The vegetation formation of the region is identified by Krajina (1959) as a part of the Coastal Western Hemlock Zone of the Pacific Northwest.

---

[11]http://maps.google.com/maps/ms?ie=UTF&msa=0&msid=2166263773098453 13321.00049bf3254ab94ea225e

On the energy-based entropy structure of natural vegetation

A common characteristic of floodplains is the presence of more or less flat levels variously identified as benches (when still flooded) or terraces (when out of reach by normal floods). Levels are formed by the joint effect of sedimentation and erosion of the flood waters. Natural floodplain levels have a slight incline toward the adjacent higher level.

The surveyed transect crosses three natural levels with elevation increasing as distance from the river increases. The average elevation ranges from 4.2 m to 5.4 m and 10.8 m above the water level of the Coquihalla River on the first day of the survey. The vegetation cover is remarkably different by appearance on the different levels and uniform on the same level[12].

Data set

The original survey gave us a data set of cover/abundance (C/A) estimates for nearly 100 species in 45 sample plots 10m x 10m each. A subsample of 17 species are shown in Table 1. The chosen species occur on all levels. We opted for the drastic reduction of the data set to avoid overwhelming the example with information we do not need for making our point.

R1, R2 and R3 designate the three metacommunities. For these the grand totals and hierarchical levels (coded for Class, Order, Family, Genus, and Species) are found on the base line of Table 1. The parameter n take on values depending on the data elements contributing to the relevant value of T in any number of different designs. The analysis follows the example in the previous chapter.

Table 1. The reduced Coquihalla data set. Brief description of site and sampling design appears in the main text. Plant identification follows standard field manuals.[13] Table headings: # -- sequence number in M. Mihály's original records; CD –

---

[12] This is an indication of different overflow frequency and duration, and the quality and quantity of the sediment load carried by the flood waters.

[13] Cronquist, C.L.A. , Owenbey, M. and J.W. Thompson. 1955-1959. Vascular plants of the Pacific Northwest. University of Washington Press, Seattle, Wash-

code vectors identifying species mappings in evolutionary plant systematics; a, b, c – floodplain levels low to high. The *C/A* totals (third segment in the table) are based on 14, 20 and 11 sample plots laid on levels a, b, c.

#	Plant taxa (species)	Class	CD	Order	CD
3	Acer macrophyllum (shrub)	Eudicots	3	Sapindales	11
60	Symphoricarpos albus	Asterids	1	Dipsacales	3
43	Polystichum munitum	Pteridopsida	5	Denntaedtiales	2
37	Mniumspinulosum	Bryopsida	2	Eubrya	5
54	Rubus spectabilis	Magnoliopsida	4	Rosales	10
65	Trientalis latifolia	Eudicots	3	Ericales	4
49	Rhytidiadelphus loreus	Bryopsida	2	Eubrya	6
34	Mahonia nervosa	Magnoliopsida	4	Ranunculales	9
24	Eurynchium oreganum	Bryopsida	2	Eubrya	6
8	Amelanchier florida	Magnoliopsida	4	Rosales	10
52	Rosa gymnocarpa	Magnoliopsida	4	Rosales	10
29	Hylocomium splendence	Bryopsida	2	Eubrya	7
40	Pachistima myrsinites	Eudicots	3	Celasrales	1
43	Pseudotsuga menziezii	Pinopsida	5	Pinales	8
3	Acer circinatum	Eudicots	3	Sapindales	11
6	Achlys triphylla	Magnoliopsida	4	Ranunculales	9
63	Thuja plicata	Pinopsida	5	Pinales	8
	Number of states		5		11

#		Family		Genus	
3	Acer macrophyllum (shrub)	Sapindaceae	1	Acer	1
60	Symphoricarpos albus	Caprifoliaceae	10	Symphoricarpos	14
43	Polystichum munitum	Dennstaedtiaceae	7	Polystichum	9
37	Mniumspinulosum	Mniaceae	5	Mnium	7
54	Rubus spectabilis	Rosaceae	2	Rubus	13
65	Trientalis latifolia	Myrsinaceae	4	Trientalis	16
49	Rhytidiadelphus loreus	Hypnaceae	6	Rhytidiadelphus	11
34	Mahonia nervosa	Berberidaceae	11	Mahonia	6
24	Eurynchium oreganum	Hypnaceae	6	Eurynchium	4
8	Amelanchier florida	Rosaceae	2	Amelanchier	3
52	Rosa gymnocarpa	Rosaceae	2	Rosa	12
29	Hylocomium splendence	Hypnaceae	6	Hylocomium	5
40	Pachistima myrsinites	Celastraceae	9	Pachistima	8
43	Pseudotsuga menziezii	Pinaceae	3	Pseudotsuga	10
3	Acer circinatum	Sapindaceae	1	Acer	1
6	Achlys triphylla	Berberidaceae	11	Achlys	2
63	Thuja plicata	Cupressaceae	8	Thuja	15
	Number of states		11		16

Species	Total a	Total b	Total c
Acer macrophyllum (shrub)	23	3	5
Symphoricarpos albus	70	18	4
Polystichum munitum	105	47	1
Mniumspinulosum	69	17	2

ington.   Grout, A.J. 1928-1940. Moss flora of North America north of Mexico. Newfane, Vermont.

Rubus spectabilis	8	43	16
Trientalis latifolia	7	38	19
Rhytidiadelphus loreus	25	47	34
Mahonia nervosa	37	132	43
Eurynchium oreganum	42	99	51
Amelanchier florida	3	8	16
Rosa gymnocarpa	2	8	15
Hylocomium splendence	23	123	73
Pachistima myrsinites	4	65	72
Pseudotsuga menziezii	114	121	88
Acer circinatum	106	105	29
Achlys triphylla	63	94	17
Thuja plicata	38	122	52
T	739	1090	537

Figure 2. Google map of the survey site on the Coquihalla floodplain at Hope, British Columbia. The sampling site (a 40 meter strip on two sides of the heavy line) and sampling details are described in the main text and in the caption of Table 1. The flow in the uncontrolled state of the river could be torrential, several meters higher than on day of sampling. The uncontrolled flow from October to March is a mere average 2 m^3/s. The flow can reach 12 m^3/s in May and June following runoff after warm days or rain by a roughly 24 hr lag. The river is fed by runoff from the 740 km^2 watershed in the Cascade Mountains. The highest peak in the watershed is around 2000 m.

Energy-based entropy numerics

The data set is in Table 1. Regarding basic parameters we should note:

1. *T = 739, 1090, 537* for the three metacommunities; *T* is the same for each hierarchical level within a given metacommunity.

2. $n=17, 5, 11, 11, 16$ in the basic analyses; take the 3x multiple of these in catenation; or 17, 3, 3x17 in the rectangular matrix.

We introduced the computational techniques in the previous chapter. We now reiterate that H_{spec} is specific entropy and it depends on the hierarchic level. For example on the Class level H_{spec} is the same as H_n. This is not that way on the lower levels. For example on the Order level we have

$$H_{n\,spec,oder} = H_{n\,oder} - H_{n\,class}$$

We pass to H_{spec} this way:

$$H_{spec,order} = \frac{H_{n\,spec,order}}{n_{order} - n_{class}}$$

Note that n_{order} is the number of nodes on the Order level and n_{class} is the number of nodes on the Class level. Further,

$$P = e^{-H_{spec}}$$

The results are presented in Tables 2, 3 and 4, and Figure 3.

Table 2. Values of H_{spec} (top five rows) and H values for the baseline data vectors (last row) for the metacommunities and catena.

H_{spec}	R1	R2	R3	Catena [R1 R2 R3]	P
Class	5.999	6.387	5.681	6.064	0.002
Order	4.561	4.946	4.246	4.625	0.010
Family	4.179	4.562	3.865	4.243	0.014
Genus	3.988	4.371	3.676	4.052	0.017
Species	3.824	4.206	3.513	3.888	0.020
Baseline H	4.783	4.468	4.468	4.848	0.008

Table 3. Analysis of the species x relevé array.

H species x relevé	H_n	H	P	H_n%
main effect (phylogeny)	100.969	5.939	0.003	40.838
main effect (gradient)	23.013	7.671	0.000	9.308
joint effect	247.239	4.848	0.008	100.000
interaction	123.258	3.976	0.019	49.854

On the energy-based entropy structure of natural vegetation

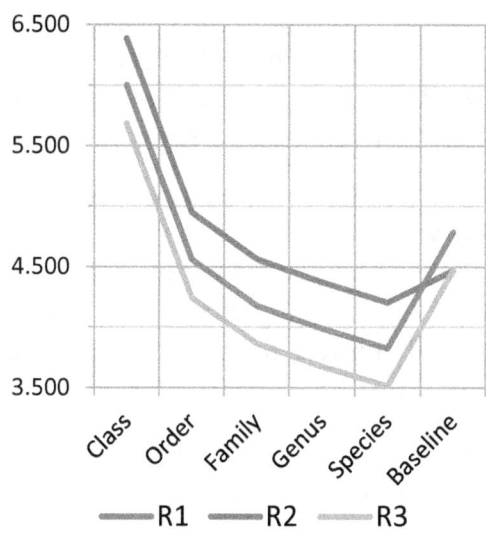

Fig. 3 Graphs of the column entries of Table 2.

What can we read from the numbers:

1. The potential energy-based entropy H_{spec} is typically on a declining path from Class to Species (see Figure 3). The upturn at Baseline is not part of the trend. It is owing to the baseline H values. If we take the difference of the highest and lowest H_{spec} value for each of the three metacommunities in Table 2 we have uniformly a rounded 2.2. The corresponding P value is 0.111. From this we conclude that the separation of the Class level from the Species level is reasonably unique.

2. On the transect the elevation incline is sharp. Is there any parallel trend in the three H_{spec} values? The answer comes from Figure 3:

a. The H_{spec} values decline from R2 (the mesic metacommunity) in both directions on the elevation gradient.

b. The greatest difference occurs on the "species" level, 4.206 - 3.513 = 0.693, a value which has associate probability P = 0.500. This is far too great to consider the greatest difference unique.

c. We conclude that the separation of the R1, R2, R3 line graphs in Figure 3 lack statistical uniqueness, therefore the elevation effect is not unique.

3. An interesting fact is that H_{spec} in the concatenated relevés is the average of the H_{spec} values in the individual relevés (see the forth column in Table 2). What do we see in the last column of Table 2? – very low probabilities. All the values are highly unique.

4. There are other aspects of the energy-based entropy structure that come to light when we analyse the rectangular matrix of the baseline data (Table 1). The question of unequal sample size comes up as a possible impediment. My answer is negative. The unequal sample size does not matter, the analysis is based on the basic quantities

$$T= 2366 \quad n_{species} = 17 \quad n_{metacom} = 3 \quad n_{joint} = 51$$

Table 3 displays the results. Before interpretations the reader should consider the basic equation:

$$C_n = \frac{(T+n)^{(T+n)}}{T^T n^n} \qquad H_n = \ln C \qquad H = \frac{H_n}{n}$$

For main effect "species" n=17, for main effect "metacommunity" n=3, and for joint effect n=51. So we have

$$C_{species}, \ C_{metacom}, \ C_{joint}$$

We have for the interaction

$$C_{interaction} = \frac{C_{joint}}{C_{species} C_{metacom}}$$

The interpretation of Table 3 is rather straight forward:

1. Based on percentages of H_n the leading carrier of energy-based entropy is "interaction" (elevation x relevés). This is followed closely by "species richness" (phylogeny). The gradient effect (relevés only) lags far behind. The P values are very low, well within any usual limits of statistical uniqueness.

2. We considered the energy-based entropy separation of meta-communities R1, R2, R3 based on the results in Table 2 and Figure 3, and the effect of the elevation gradient on metacommunity separation, not unique in energy terms.

We examined the gradient itself and found the associated main effect (Table 3) small numerically, yet statistically highly significant. Is this not a contradiction? No, it is not. The separation is not unique, the total contribution associated with the gradient, particularly if we consider the interaction term as well, is unique.

QUESTIONS AND ANSWERS

The interpretation of computed energy-based entropy value should begin with the recollection that T and n are the quantities in the computation of C and in turn for the determination of H. This emphasizes the dependence of the outcome upon taxon richness (n) and relevé total T. I tried to anticipate questions regarding the exercise:

1. Why did we have to assume that energy comes in discrete units? The proposition that energy comes in discrete, countable units has been the main point in Max Planck's 2001 seminal paper. The assumption is needed for valid application of the equation

$$H_n = k \ \ln \ C + constant$$

2. What justifies calling H_n energy-based entropy? This is another point Max Planck makes in his 2001 paper. He gives as the proof in the manner of

On the energy-based entropy structure of natural vegetation

$$H_n = k \ln C = k \ln \frac{(n+T)^{n+T}}{n^n T^T} = kn\left[\left(1+\frac{U}{\varepsilon}\right)\ln\left(1+\frac{U}{\varepsilon}\right)-\frac{U}{\varepsilon}\ln\frac{U}{\varepsilon}\right]$$

3. Where is the probabilistic connection of H_n? When we assume discrete, countable energy units, and assume further the rule of chance over the realisation of an observed complex (hierarchical relevé, metacommunity) with a given n and a given T, the complex we actually have is one of

$$C = \frac{(n+T)^{n+T}}{n^n T^T}$$

equally probable complexes. Each has an equal chance of material-ising. Therefore the probability of the hierarchical relevé as we observed it has a probability of occurring by chance exactly equal to $\frac{1}{C}$. When this has a small value (0 to 1 scale), the relevé is considered unique. When it has a large value, the relevé is considered common.

4. We accepted from Max Planck that H_n is a proxy parameter for energy. What kind of energy? The energy implied is potential energy.

4. What is the reason for using hierarchical relevés? The hierarchical relevé puts a handle on the problem of interconnecting the principle factors which account for directed variation in n and T. In our examples, one of these is phylogeny and the other current environmental mediation.

5. Are taxa and environmental variables not left out of the conclusions when we apply the methodology just described? True, the conclusions we draw are at least one step removed from the community elements and also from the environmental variables. Therefore any comparison made between complexes (hierarchical relevés or levels in hierarchical relevés) is a comparison of their energy-based entropy states. In this regard the methodology joins the group of information theoretical techniques whose purpose is the interrogation of the community's diversity structure. By so do-

ing the discovery of community assembly rules is in such terms.

6. What is the ultimate statement we can make after completion of the analysis? In the briefest, it is a probabilistic statement regarding the energy structure of the metacommunity, a condition of phylogeny and environmental mediation.

REFERENCE BIBLIOGRAPHY

Bentham, G. and J.D. Hooker. 1862–1883. Genera plantarum ad exemplaria imprimis in herbariis kewensibus servata definita. 3 volumes. Biodiversity Heritage Library,
http://www.biodiversitylibrary.org/item/14680

Bio, A.M.F., Alkemande, R., and A. Barendregt. 1998. Determining alternative models for vegetation response analysis: a non-parametric approach. Journal of Vegetation Science 9: 5-16.

Camazine, S., Deneubourg, J.L., Franks, N.R. Sneyd, J., Theraulaz G. and E. Bonabeau. 2003. Self-Organization in Biological Systems, Princeton University Press.

Diaz, S., Acosta, A., and M. Cabido. 1994. Grazing and the phenology of flowering and fruiting in a montane grassland in Argentina – a niche approach. Oikos 70: 287-295.

Dobzhansky, T. 1937. Genetics and the Origin of Species. Columbia University Press.

Felsenstein, J. 2004. Inferring Phylogenies. Sinauer Associates, Sunderland, MA.

Greig-Smith, P. 1952. The use of random and contiguous quadrats in the study of the structure of plant communities. Annals of Botany 16: 293-316.

Henning, W. 1965. "Phylogenetic Systematics," Ann. Rev. Entomol., Vol. 10, 97-116

Huxley, J. 1942. Evolution: the modern synthesis. The MIT Press.

Kerner von Marilaun, A. 1863. Das Pflanzenleben der Danauländer. Innbruck, Wagner.

Krajina, V.J. 1959. Bioclimatic Zones in British Columbia. UBC Botanical Series #1, Vancouver, B.C.

Mayr, E. and W.B. Provine (eds). 1998. The Evolutionary Synthesis: Perspectives on the Unification of Biology. Harvard University Press.

Mayr, E. 2002. What evolution is. Weidenfeld & Nicolson, London.

Orlóci, L. 1965. The Coastal Western Hemlock Zne on the southwestern British Columbia Mainland. Vegetation-environmental patterns and ecosystem classification. In: V.J. Krajina (ed.), Ecology of Western North America. Vol. 1, pp. 18-34.

Orlóci, L. 1971. An information theory model for pattern analysis. Journal of Ecology 59:343-349.

Orlóci, L. 1991. On character-based community analysis: choice, arrangement, comparison. In: Feoli, E. and L. Orlóci (eds.), Computer Assisted Vegetation Analysis, pp. 81-93. Kluwer Academic Publishers, London.

Orlóci, L. 2006. Diversity partitions in 3-way sorting: functions, Venn diagram mappings, typical additive series, and examples. Community Ecology 7:253-259.

Orlóci, L. 2009. Multi-scale trajectory analysis: powerful conceptual tool for understanding ecological change. Frontiers of Biology in China 4: 158-179

Orlóci, L. 2010. Statistical Ecology: a reasoned approach. SSCADA Publishing. Internet Edition

Orlóci, L. and M. Orlóci. 1985. Comparison of communities without the use of species: model and example. Ann. Bot. (Roma) 43:275-285.

Pillar, V.D. and L.S. Duarte. 2010. A framework for metacommunity analysis of phylogenetic structure. Ecology Letters 13: 587–596.

Pillar, De Patta V. and L. Orlóci. 1993. Character-based Vegetation Analysis: the Theory and an Application Program. Ecological Computations Series (ECS): Vol. 5. SPB Academic Publishing bv, The Hague, The Netherlands.

Planck, Max. 1901. On the law of distribution of energy in the normal spectrum. Annalen der Physik. Vol. 4, p. 553 et seq.

Podani, J. 2003. The Evolution and Systematics of Terrestrial Plants. In Magyar. Elte Ötvös Kiadó, Budapest.

Podani, J. 2010. Taxonomy in Evolutionary Perspective. An essay on the relationships between taxonomy and evolutionary theory. Synbiologia Hungarica 6:1-42.

Podani, J. 2012. Tree thinking, time and topology: comments on the interpretation of tree diagrams in evolutionary/phylogenetic systematics. Cladistics 1–13, Wiley-Blackwell.

Rényi, A. 1961. On measures of entropy and information. In: J. Neyman (ed.), Proceedings of the 4th Berkeley Symposium on Mathematical Statistics and Probability, pp. 547-561. University of California Press, Berkeley.

Revell, L.J., Harmon, L.J. and D.C. Collar. 2008. Phylogenetic Signal, Evolutionary Process, and Rate. Oxford Journals, Life Sciences, Systematic Biology Volume57, Issue 4, Pp. 591-601.

Schuh, R.T. and A.V.Z. Brower. 2009. Biological Systematics: Principles and Applications. (2nd edn.) Cornell University Press .

Singh, G. 2004. Plant Systematics: An Integrated Approach. Science Publishers, Enfield, NH. ISBN 978-1-57808-351-0

Stachowicz, J.J. 2001. Mutualism, facilitation, and the structure of ecological communities. BioScience 51: 235-246.

Stebbins, G. L. 1950. Variation and Evolution in Plants. Columbia University Press, New York.

Sukopp, H. 1987. On the history of plant geography and plant ecology in Berlin. Englera 7: 85-103.

Tilman, D. 2004. Niche tradeoffs, neutrality, and community structure: A stochastic theory of resource competition, invasion, and community assembly. PNAS July 27, 2004 vol. 101, no. 30 10854-10861.

Warming, E. with M. Vahl. 1909. Oecology of Plants - an introduction to the study of plant-communities translated by P. Groom and I. B. Balfour. Clarendon Press, Oxford. ---- Original: Warming, E. 1895. *Plantesamfund - Grundtræk af den økologiske Plantegeografi*. P.G. Philipsens Forlag, Kjøbenhavn.

Wildi, O. and M Schütz. 2000. Reconstruction of a 405 yr. recovery process from pasture to forest Community Ecology 1: 25-32.

Wilson, J.B. 2009. Assembly rules in plant communities, pp.130-164. In: E. Weher and P. Keddy (eds.), Ecological Assembly Rules, Cambridge Books Online.

http://ebooks.cambridge.org/chapter.jsf?bid=CBO9780511542237&cid=CBO9780511542237A013

Wilson, J.B., Ulman, I. and P. Bannister. 1996. Do species assemblages recur? Journal of Ecology 84: 471-474.

INDEX

László Orlóci

READER'S NOTES

BIOGRAPHIC NOTES

 László Orlóci was born in Hungary in 1932. He holds degrees in forest engineering (DFE Sopron), forestry science and biology (BSF, MSc, PhD University of British Columbia), and DSc *h.c.* in biology (University of Trieste). Orlóci held appointments as NATO Science Fellow (University College of North Wales), professor (University of Western Ontario), and visiting professor at universities in the Americas, the Pacific, Asia, and Europe.

Orlóci is an INTECOL Distinguished Statistical Ecologist. He is external (academician) Member of the Hungarian Academy of Sciences, and regular (academician) Fellow of the Academy of Sciences of the Royal Society of Canada.

Orlóci published over 100 papers in scientific journals, numerous monographs, books and book chapters. His current essays on trajectory analysis, the rules of process governance, and the phylogenetic signal in vegetation transitions have considerable significance for evolutionary ecology and global change science. His present work on energy structures in metacommunities is seminal, pointing to a new direction.

Orlóci is married to author Márta Mihály, Sopron forest engineering alumna. Their daughter Martha Barbara is a Geography graduate of UWO.

www.ingramcontent.com/pod-product-compliance
Lightning Source LLC
Chambersburg PA
CBHW051257170526
45165CB00004B/1755